**NATIONAL GEOGRAPHIC**

**School Publishing**

# The Sun

## PATHFINDER EDITION

By Fran Downey

## CONTENTS

# The S

**By Fran Downey**

Our sun resembles a large yellow ball in the sky. It always looks peaceful and calm. It never seems to change. But looks can be deceiving. "It's not a boring white disk," says Bernhard Fleck. He studies the sun.

In fact, the sun may be the most active object in our solar system. Its surface is a bubbling kettle of superhot gases. The sun fires streams of these gases into space. Some even collide with Earth.

Clearly, there's a lot to learn about our stormy star. Let's start with exactly what it is.

## A Real Star

In a word, our sun is a **star**. That means it's a huge ball of glowing gases. The sun is larger than all the planets and moons in our solar system combined. A million planets the size of Earth could fit inside it!

Size isn't everything, though. The really hot thing about a star is that it creates light and heat. Planets and moons can't do that. Light from the sun warms all the planets in our solar system. Without it, life on Earth would be impossible.

## Star Studies

The sun looks different from any other star we can see. That's because it is only about 150 million kilometers (almost 93 million miles) away. In space, that's just around the corner. The next closest star is over nine and a half trillion kilometers (about six trillion miles) from Earth. All the other stars are much farther away.

Aside from that, our sun is an average star. **Astronomers**, or space scientists, have found millions like it. Studying our sun, then, is a way to learn more about countless other stars too. For scientists, our sun is a huge lab.

**Solar Express.** Explosions in the sun's atmosphere shoot gases into space. These may shoot 321,869 kilometers (200,000 miles) across the surface of the sun.

## Hot Stuff

The sun is made largely of hydrogen. It also contains some helium. Both are gases.

In the center of the sun, the temperature soars to almost 15 million degrees Celsius (°C) [27 million degrees Fahrenheit (°F)]. At that temperature, hydrogen is turned into helium. That releases heat and light.

Sunbeams made in the core, or center, have a long journey ahead of them. Material in the core is jammed together tightly. Light bounces off this material. After bouncing around for hundreds of thousands of years, it finally reaches the surface, or **photosphere**.

Once light breaks through the surface, it has an easier time. Light moves away from the sun at 299,338 kilometers (186,000 miles) a second. It reaches Earth in about eight minutes.

## Color Coded

Picture the night sky. Most of the stars look like white dots. Yet stars actually come in different colors. Knowing the color of a star helps you tell how old and how hot it is.

**Blue stars** are babies. They're usually not much more than a billion years old. Blue stars are also the hottest. The surface of a blue star can be a sizzling 11,093°C (20,000°F).

**Yellow stars** are grown-ups. They've been around for several billion years. Over time, stars grow cooler. So a yellow star is only about 5,538°C (10,000°F) on its surface.

**Red stars** are old folks. They've had billions of birthdays. That means the surface of a red star is downright chilly. Indeed, temperatures there barely break 2,760°C (5,000°F).

Where does our sun fit into the star family? Well, it's yellow. So it's in the middle. It is not too hot or cool. It is just right for life on Earth.

**In the Loop.** You could stack ten Earths beneath these arches of superhot gas.

## Spots on the Sun

Our sun is not completely yellow, though. Pictures of its surface show dark blotches. They are called sunspots. These areas are several thousand degrees cooler than the rest of the surface. That's why they look dark.

On the sun, even a spot is huge. Small sunspots are more than a thousand miles across. The largest are bigger than Earth.

Sunspots come and go. They follow a 22-year cycle. At the start, there may be just a few sunspots. Over the next 11 years, more appear. There can be 100 spots at the cycle's peak. Then fewer spots start appearing. Eleven years later, it all begins again. We don't know why.

**Spots of Trouble?** Huge spots freckled the sun's face on October 28, 2003. Just a few days later, a mightly solar flare erupted. It was the largest ever recorded.

## Sunny Weather?

The sun's atmosphere is even hotter than its surface. In fact, the outer layer is more than 1,111,093°C (2,000,000°F).

That layer, called the **corona**, is a busy place. Gases there cause huge explosions. Just one of those blasts is more powerful than billions of atomic bombs.

Explosions in the corona sometimes make huge arches of hot gas. The gas races many thousands of miles away from the sun. Then it falls back toward the surface.

At other times, giant **flares** erupt from the sun. A flare is a bit like water rushing from a fire hose. Instead of water, though, the flare is made of light and heat.

Flares can travel far. Some even reach Earth's atmosphere. They don't make it to the planet's surface. But they can harm satellites or disrupt phone and power systems.

Fortunately, scientists are learning more and more about flares. They may even figure out how to predict them. That would help us protect our technology.

Of course, predicting flares is merely one benefit of studying the sun. Even more important is what our star can teach us about trillions of other stars.

*What are some of the ways that our sun affects Earth?*

# Wordwise

**astronomer:** scientist who studies space

**corona:** outer layer of the sun's atmosphere

**flare:** stream of light and heat

**photosphere:** surface of the sun

**star:** ball of gas in space that can make heat and light

**sunspot:** dark, cooler area on the sun's surface

**1** Flare

**Magnetic Field** **2** ⟶

**3** Core

# Inside the Sun

4 Sunspots

5 Photosphere

6 Chromosphere

7 Corona

1 **Flare:** giant explosion that makes a powerful stream of energy

2 **Magnetic Field:** invisible electric currents that act like huge magnets (The large loops seen on the surface are part of the field.)

3 **Core:** huge furnace in center of the sun that makes heat and light

4 **Sunspots:** dark areas that are cooler than the rest of the sun's surface

5 **Photosphere:** visible surface that we see from Earth

6 **Chromosphere:** inner layer of the sun's atmosphere

7 **Corona:** outer layer of the sun's atmosphere

# Earth

AND THE **Sun**

The sun may be the hottest topic in our solar system. But if you're like most people, you probably don't give it much thought. It rises, and it sets—end of story. Well, think again. Energy from the sun makes life on Earth possible,

## Heating Up Earth

Without the sun, Earth would be a very different place. There would be no plants, no animals, and no people. Our planet would be a frozen chunk of rock spinning through space.

Luckily, we have the sun. It warms Earth. You can feel the sun's heat on a hot summer day. Yet even in winter, the sun gives off heat and warms our planet.

Why does the place where you live get warmer or cooler as the seasons change? It has to do with how Earth moves. Each year Earth completes one orbit, or loop, around the sun. As our planet travels, the seasons change. Areas that are tilting toward the sun are warmer. They have spring and summer. Areas that are tilting away are colder. They have fall and winter.

## Food From the Sun

The sun doesn't just warm our planet. It gives us light too. Plants and animals need the energy in sunlight to survive.

Plants are unique. They use sunlight to make food. The food gives them energy to live. Animals can't make their own food from sunlight. Yet in a way, sunlight feeds them too. How?

Many animals munch on grass or grains. These animals get energy from plants, which got energy from the sun. The same is true when animals eat other animals. Some of the animals that are eaten ate plants—and those plants depended on the sun.

So whether you like salads or meat, you are getting energy that started out as light from the sun.

**Food Energy.** Plants, such as this corn, use energy from the sun to grow. In turn, plants give animals energy to live.

## Colors of the Rainbow

Sunlight also brings color to our world. That's because sunlight is actually made of colors. Don't believe it? Just look at a rainbow.

A rainbow forms when sunlight hits rain in the air. When sunlight moves through the raindrops, a band of colors appears in the sky. You see red, orange, yellow, green, blue, indigo, and violet. These are all the colors in light.

The colors of objects, such as rocks or flowers, also come from light. When sunlight hits an object, some colors are absorbed—they soak in. Others are reflected, or bounce off.

Your eyes see the colors that bounce. Suppose you're looking at a red flower. The flower absorbs all of the colors in light except for red. The red bounces off the flower and back to your eyes. So you see the flower as red.

## Sun Power

The sun gives us colors, food, and warmth. But the list doesn't stop there. We also use the sun as a source of power.

Solar cells are machines that change sunlight into electricity. Solar cells can run calculators and toys. They can also be linked together to power cars, homes, and even satellites in space.

Solar cells are useful. Yet most of our power comes from coal, oil, and gas. These fossil fuels formed from the buried remains of ancient plants and animals. The plants and animals got their energy from the sun. So even the energy in fossil fuels started out as sunlight!

The sun affects our lives in countless ways. It's a powerful source of energy. It makes life on our planet possible.

**Colors in the Sky.**
A rainbow shows the many colors that make up sunlight.

# The Sun

**Answer these questions to find out what you learned about this hot topic.**

**1** Why does the sun look different to us than other stars?

**2** What is the sun made of?

**3** What is the difference between blue, yellow, and red stars? Which kind of star is our sun?

**4** What are flares? How can they affect people on Earth?

**5** How is the sun important to life on Earth?